innovation

Alex Ndukwe

First printing, 2020

Published by Nay publications

Printed in the United States of America

ISBN: 9798697946831

Dedication

To my friend, Engr Israel Omosagba, an innovator to the core, he manufactured 382V MPPT Charge controller.

Forward

Innovation would enable us to take a giant stride in our society, Life problems would be resolved and most importantly manpower shortage would be addressed.

We must have vision, be ready to run with it on daily basis, the book showcases technologies that are available and its application in various human endeavours.

The future technologies were discussed in chapter 3, we can see wastes turned to energy, coffee bean converted to power. Visualize, ponder on it then you will realize that there is a possibility.

Innovations would proffer solutions to problems with ease, I want to encourage us to be creative and refuse mediocrity.

Alex Ndukwe

Table of Contents

Conclusions

Innovation in perspective

Automated Processes has improved life and business all over the world, shortage of professional manpower in various fields of human endeavour propelled such ideas. We tend to hear of various terminologies like Internet of Things (IOT), Robotics, Machine Learning, Artificial Intelligence etc. These are ideas to make life easy and bearing in mind that the focus to ease stress experienced by individuals, corporations, and other entities.

Africa is behind other continents, though we might attribute this to other issues like corruption, bad Leadership, poverty, and the likes but this is not the focus of this book. I am not shying away from the truth; these issues must be fixed before we can see light at the end of the tunnel. Policies, frameworks are the things that would make it possible for embrace, implementation, execution of proposed or existing innovation. The enabling environment must be present for the perceived solutions to thrive.

It's important to distinguish between inventions and innovations. Invention is a subset of innovation. The power of innovation lies in new value creation on a commercial scale. When an invention is exploited successfully commercially, it becomes an innovation. For instance, the electric bulb is a great invention, but producing them in bulk, and serving millions of customers is what made it an innovation.

Innovation also can be defined as something that adds value to what you are already doing, in a unique, unprecedented way which has the potential to add value to the community/ stakeholders.

There must be a synergy of all stakeholders for results to be achieved. Innovators cannot work in silos; the handshake must be in existence for the impact to be felt. Let me share this, the number of doctors in Nigeria is **forty-five thousand**, this cannot match the population of **two hundred**

million. Automation is required to breach the gap and ensure that an improvement in health care service is achieved, it has become fashionable to see medical Application with artificial intelligence been deployed by corporate institutions with the sole aim of breaching the gap, but the question here is how effective has this effort been coupled with the fact that the Ministry of health might not be involved, meaning there is no framework that supports such deployments. It is bound to fail at the end of the day, the corporate bodies should do the needful. Rural communities should be adequately covered, though we might claim health centres are already in such communities, there are no issues, but technology should be deployed to complement the efforts of the doctors.

Leadership in Africa should arise in this direction, when it comes to tax collection , the vigour and energy can be seen or felt , 10% of this energy should be deployed to technology

innovations for sectors of the economy like education, Agriculture, healthcare, etc.

providing an enabling environment for these deployment to thrive.

Technology will lift our world, though fears will arise that job cuts are eminent if this happens. Can we imagine how things would be without innovative ideas in solving problems, a disaster.

Chapter One

Roadmap to success

Automation will improve quality of service and improve turnaround time with respect to processes. I want us to visualize the banking environment, Applications has reshaped how customers are served and most importantly led to emergence of **digital banks** around us today, you do not need to visit banking halls. Emergence of virtual halls has led to shutdown of physical bank branches; the wisdom is that they have been reduced to cut operational cost that would have negative effect on the bottom-line. No one anticipated COVID19 and the effect it has had on business, earnings are very lean compared to operational cost.

Most economies in Africa have embraced innovative trends on a minimal scale, the major challenge is the cost. We can imagine the cost of some robots, very expensive and cost a fortune to acquire, but we must look at sectors that are very vital to the economy, a classical example is Agriculture, automation promises improvements with respect to high yields, data collection , improved processes etc. I am quite aware that it might sound absurd to some of us but embracing these changes, would improve production and make Africa food basket of the world.

Applications in our work environment would also aid improvements of services, Financial institutions have leveraged on this, they spend heavily on automation to a larger extent than any sector in any country, the essence is that customers should enjoy banking services from the comfort of their homes 24/7. Public sector should take a cue from this and ensure processes are automated to improve service.

We should also note that there might be some limitations, let us look at this example, 'deploying drones to supplement efforts of troops during a war', might require obtaining composite approvals from the government and also ensuring availability of Legal framework that supports such operation. This implies that stakeholders must be involved before it can see the light of the day.

Risks associated with innovations must be considered, seeking for ways to mitigate such risks, deployment of Applications must be protected by various IT security applications to ensure that the users are protected against financial loss, providing tit bits for users so as to be aware of cyber criminals. Financial institutions are doing good job in this regard. Efforts must be in top gear to protects Technology infrastructures like servers, network environments, databases, devices etc. Preventing hackers from having access to them.

- Releasing malwares to devices in a bid to make them malfunction.

Most importantly associated cost of deployment of such infrastructure should be weighed vis a vis its benefit to the organisation, it must be a worthwhile investment with reasonable cost, it shouldn't throw the organization off balance, it should rather improve earnings.

The people that will deploy and manage the technology, are they well trained and equipped for challenges ahead. Obviously, we need skilled people to handle issues associated with the said innovations, this would guarantee a hitch free deployment and desired benefits would be derived.

The next chapter would afford us the opportunity to discuss some innovations that are already available and proposed solutions to some problems around us today, efforts have been made to cover nearly most areas of human endeavours.

"Anything's possible if you've got enough nerve." –J.K. Rowling

Chapter two

Automation of Processes:

This chapter is the core of this book, we will discuss innovation with respect to the following sectors like **Agriculture**, **Healthcare**, **Manufacturing**, **Power**, **Banking and Finance**, **Procurement**, **Warehouse**, **Information Systems**, **Logistics** and **Physical Security**.

Agriculture

This is a very vital sector in any country's economy, food production that would cater for their needs and export to other countries to earn foreign exchange. Essence of technology is to increase the yield, produce food round the year not minding the season, easy data collection, improving soil quality.

Innovation is more important in modern agriculture than ever before. The industry is facing huge challenges, from rising costs of supplies, a

"If you want something new, you have to stop doing something old." – Peter F. Drucker

shortage of labour, and changes in consumer preferences for transparency and sustainability. There is increasing recognition from agriculture corporations that solutions are needed for these challenges. In the last 10 years, agriculture technology has seen a huge growth in investment. Major technology innovations in the space have focused around areas such as **indoor vertical farming**, **automation** and **robotics**, **livestock technology**, **modern greenhouse practices**, **precision agriculture** and **artificial intelligence**, and **blockchain**.

Indoor Vertical Farming

Indoor vertical farming can increase crop yields, overcome limited land area, and even reduce farming's impact on the environment by cutting down distance travelled in the supply chain. Indoor vertical farming can be defined as the practice of growing produce stacked one above another in a closed and controlled

"Anything's possible if you've got enough nerve." –J.K. Rowling

environment. By using growing shelves mounted vertically, it significantly reduces the amount of land space needed to grow plants compared to traditional farming methods. This type of growing is often associated with city and urban farming because of its ability to thrive in limited space. Vertical farms are unique in that some setups do not require soil for plants to grow. Most are either hydroponic, where vegetables are grown in a nutrient-dense bowl of water, or aeroponic, where the plant roots are systematically sprayed with water and nutrients. In lieu of natural sunlight, artificial grow lights are used. Vertical farms use up to 70% less water than traditional farms.

From sustainable urban growth to maximizing crop yield with reduced labour costs, the advantages of indoor vertical farming are apparent. Vertical farming can control variables such as light, humidity, and water to precisely measure year-round, increasing food production with reliable harvests. The reduced water and energy usage optimize energy conservation --

"If you want something new, you have to stop doing something old." – Peter F. Drucker

vertical farms use up to 70% less water than traditional farms. Labour is also greatly reduced by using robots to handle harvesting, planting, and logistics, solving the challenge farms face from the current labour shortage in the agriculture industry.

Farm Automation

Farm automation, often associated with "smart farming", is technology that makes farms more efficient and automates the crop or livestock production cycle. An increasing number of companies are working on robotics innovation to develop drones, autonomous tractors, robotic harvesters, automatic watering, and seeding robots. Although these technologies are new, the industry has seen an increasing number of traditional agriculture companies adopt farm automation into their processes. New advancements in technologies ranging from robotics and drones to computer vision software have completely transformed modern agriculture. The primary goal of farm automation technology is to cover easier, mundane tasks. Some major technologies that are most commonly being utilized by farms include: harvest automation, autonomous tractors, seeding and weeding, and drones. Farm automation technology addresses major issues like a rising global population, farm labor shortages, and changing consumer preferences. The benefits of automating traditional farming processes are monumental by tackling issues from consumer preferences, labor shortages, and the environmental footprint of farming.

Livestock Farming Technology

The traditional livestock industry is a sector that is widely overlooked and under-serviced, although it is arguably the most vital. Livestock provides much needed renewable, natural resources that we rely on every day. Livestock management has traditionally been known as running the business of poultry farms,

"Anything's possible if you've got enough nerve." –J.K. Rowling

dairy farms, cattle ranches, or other livestock-related agribusinesses. Livestock managers must keep accurate financial records, supervise workers, and ensure proper care and feeding of animals. However, recent trends have proven that technology is revolutionizing the world of livestock management. New developments in the past 8-10 years have made huge improvements to the industry that make tracking and managing livestock much easier and data-driven. This technology can come in the form of nutritional technologies, genetics, digital technology, and more.

Livestock technology can enhance or improve the productivity capacity, welfare, or management of animals and livestock.

Livestock technology can enhance or improve the productivity capacity, welfare, or management of animals and livestock. The concept of the 'connected cow' is a result of more and more dairy herds being fitted with

"If you want something new, you have to stop doing something old." – Peter F. Drucker

sensors to monitor health and increase productivity. Putting individual wearable sensors on cattle can keep track of daily activity and health-related issues while providing data-driven insights for the entire herd. All this data generated is also being turned into meaningful, actionable insights where producers can look quickly and easily to make quick management decisions.

Animal genomics can be defined as the study of looking at the entire gene landscape of a living animal and how they interact with each other to influence the animal's growth and development. Genomics help livestock producers understand the genetic risk of their herds and determine the future profitability of their livestock. By being strategic with animal selection and breeding decisions, cattle genomics allows producers to optimize profitability and yields of livestock herds.

"Anything's possible if you've got enough nerve." –J.K. Rowling

Sensor and data technologies have huge benefits for the current livestock industry. It can improve the productivity and welfare of livestock by detecting sick animals and intelligently recognizing room for improvement. Computer vision allows us to have all sorts of unbiased data that will get summarized into meaningful, actionable insights. Data-driven decision-making leads to better, more efficient, and timely decisions that will advance the productivity of livestock herds.

Modern Greenhouses

In recent decades, the Greenhouse industry has been transforming from small scale facilities used primarily for research and aesthetic purposes (i.e., botanic gardens) to significantly more large-scale facilities that compete directly with land-based conventional food production. Combined, the entire global greenhouse market currently produces nearly US $350 billion in vegetables annually, of which U.S. production

"If you want something new, you have to stop doing something old." – Peter F. Drucker

comprises less than one percent. The entire global greenhouse market currently produces nearly Naira 145 trillion in vegetables annually.

As the market has grown dramatically, it has also experienced clear trends in recent years. Modern greenhouses are becoming increasingly tech-heavy, using LED lights and automated control systems to perfectly tailor the growing environment. Successful greenhouse companies are scaling significantly and located their growing facilities near urban hubs to capitalize on the ever-increasing demand for local food, no matter the season. To accomplish these feats, the greenhouse industry is also becoming increasingly capital-infused, using venture funding and other sources to build out the infrastructure necessary to compete in the current market.

Nowadays, in large part due to the tremendous recent improvements in growing technology, the industry is witnessing a blossoming like no time before. Greenhouses today are increasingly emerging that are large-scale, capital-infused, and urban-cantered.

Precision Agriculture

Agriculture is undergoing an evolution - technology is becoming an indispensable part of every commercial farm. New precision agriculture companies are developing technologies that allow farmers to maximize yields by controlling every variable of crop farming such as moisture levels, pest stress, soil conditions, and micro-climates. By providing more accurate techniques for planting and growing crops, precision agriculture enables farmers to increase efficiency and manage costs.

Precision agriculture companies have found a huge opportunity to grow. The emerging new generation of farmers are attracted to faster, more flexible start-ups that systematically maximize crop yields.

"Anything's possible if you've got enough nerve." –J.K. Rowling

Blockchain

Blockchain's capability of tracking ownership records and tamper-resistance can be used to solve urgent issues such as food fraud, safety recalls, supply chain inefficiency and food traceability in the current food system. Blockchain's unique decentralized structure ensures verified products and practices to create a market for premium products with transparency.

Food traceability has been at the centre of recent food safety discussions, particularly with new advancements in blockchain applications. Due to the nature of perishable food, the food industry at whole is extremely vulnerable to making mistakes that would ultimately affect human lives. When foodborne diseases threaten public health, the first step to root-cause analysis is to track down the source of contamination and there is no tolerance for uncertainty.

"If you want something new, you have to stop doing something old." – Peter F. Drucker

Blockchain can be used to solve urgent issues such as food fraud, safety recalls, supply chain inefficiency and food traceability in the current food system. Consequently, traceability is critical for the food supply chain. The current communication framework within the food ecosystem makes traceability a time-consuming task since some involved parties are still tracking information on paper. The structure of blockchain ensures that each player along the food value chain would generate and securely share data points to create an accountable and traceable system. Vast data points with labels that clarify ownership can be recorded promptly without any alteration. As a result, the record of a food item's journey, from farm to table, is available to monitor in real-time.

The use cases of blockchain in food go beyond ensuring food safety. It also adds value to the current market by establishing a ledger in the

"Anything's possible if you've got enough nerve." –J.K. Rowling

network and balancing market pricing. The traditional price mechanism for buying and selling relies on judgments of the involved players, rather than the information provided by the entire value chain. Giving access to data would create a holistic picture of the supply and demand. The blockchain application for trades might revolutionize traditional commodity trading and hedging as well. Blockchain enables verified transactions to be securely shared with every player in the food supply chain, creating a marketplace with immense transparency.

Artificial Intelligence

The rise of digital agriculture and its related technologies has opened a wealth of new data opportunities. Remote sensors, satellites, and UAVs can gather information 24 hours per day over an entire field. These can monitor plant health, soil condition, temperature, humidity, etc. The amount of data these sensors can generate is

"If you want something new, you have to stop doing something old." – Peter F. Drucker

overwhelming, and the significance of the numbers is hidden in the avalanche of that data.

The idea is to allow farmers to gain a better understanding of the situation on the ground through advanced technology (such as remote sensing) that can tell them more about their situation than they can see with the naked eye. And not just more accurately but also more quickly than seeing it walking or driving through the fields.

Remote sensors enable algorithms to interpret a field's environment as statistical data that can be understood and useful to farmers for decision-making. Algorithms process the data, adapting and learning based on the data received. The more inputs and statistical information collected, the better the algorithm will be at predicting a range of outcomes. And the aim is that farmers can use this artificial intelligence to achieve their goal of a better harvest through making better decisions in the field.

Livestock Farming Technology in Animal Agriculture

Most of us are familiar with the term "farm-to-table". We set out on sunny Saturday mornings for brunch at trendy cafés and restaurants, but while we gulp down our mimosas, we often forget the undertaking required to bring our favourite dishes from farms to kitchens. Farmers today are facing challenges from infrastructure to connectivity, growing demands for animal proteins to food spoilage, and disease with concerns rising around animal health. Technology is disrupting all industries in our modern age, and AgTech is no exception. We sought out on a mission to get back to our roots and gather perspective from those leading the industry and working to address these issues.

The traditional livestock industry is a sector that is widely overlooked and under-serviced, although it is arguably the most vital. Livestock provides much needed renewable, natural resources that we rely on every day. So why is the process to adopt technology so slow within this industry? Well the short answer, money. The economics of the livestock industry shifts from season to season, meaning it's not always profitable. With fewer investments, comes fewer adoptions of technology.

Investors need to understand the market dynamics that farmers face. "We can't view farming in the Silicon Valley mind frame since farms don't operate in the same manner," says Amado Guloy, CEO and Founder of Rex Animal Health. IoT has taken off and investments in IoT continue to increase, yet most investors don't realize that big data is often inaccessible to farmers because of lack of infrastructure and internet connectivity.

Companies such as Cainthus and Rex Animal Health are working to address some of these industry's pain points. Cainthus' cutting-edge technology monitors cows 24/7, 365 days a year, analysing their well-being, productivity, and performance. Using computer vision and artificial intelligence, they translate visual information into actionable data. In seconds, their imaging technology identifies and memorizes individual animals by their unique features. Rex Animal Health brings together clinical, performance, health, and genetic data to provide a clear understanding for farmers to prevent and predict disease in herds, that in turn can optimize yields.

Digitalization of Livestock Management

Livestock management has traditionally been known as running the business of poultry farms, dairy farms, cattle ranches, or other livestock-related agribusinesses. Livestock managers must keep accurate financial records, supervise workers, and ensure proper care and feeding of animals.

"Anything's possible if you've got enough nerve." –J.K. Rowling

However, recent trends have proven that technology is revolutionizing the world of livestock management. New developments in the past 8-10 years have made huge improvements to the industry that make tracking and managing livestock much easier and data-driven. This technology can come in the form of nutritional technologies, genetics, digital technology, and more.

Sensors are being developed to monitor real-time milk quality, health, and pregnancy hormones. In addition, virtual fences exist that can move

animals wearing a sensor to be moved remotely from one area of a pasture to another. Even robotics are advancing fast in this industry, where it's addressing the challenges of labor shortages on traditional livestock farms. 12% of dairy farms are currently using robots and is expected to grow to 20% in the next 5 years.

"If you want something new, you have to stop doing something old." – Peter F. Drucker

Livestock Technology and the 'Connected Cow'

Everything in the digital age is connected, including farming and agriculture. Livestock technology can enhance or improve the productivity capacity, welfare, or management of animals and livestock. The concept of the 'connected cow' is a result of more and more dairy herds being fitted with sensors to monitor health and increase productivity. Putting individual wearable sensors on cattle can keep track of daily activity and health-related issues while providing data-driven insights for the entire herd. All this data generated is also being turned into meaningful, actionable insights where

producers can look quickly and easily to make quick management decisions.

Cainthus is one startup that is leading innovation in the livestock technology space. Cainthus is a machine vision company ID's animal through the use of cameras and monitor their behaviors - time spent eating, drinking, and feed locations throughout the day. By monitoring feed delivery, they can deliver a body condition score that tells farmers real-time health statistics on their livestock such as weight, fat, and even if the animal has a limp. By monitoring a whole host of animal behavior, Cainthus's technology can allow producers to have better management of their animals, thus having healthier livestock with better use of resources.

What is Animal Genomics?

Animal genomics can be defined as the study of looking at the entire gene landscape of a living animal and how they interact with each other to influence the animal's growth and development. Genomics help livestock producers understand the genetic risk of their herds and determine the future profitability of their livestock. By being strategic with animal selection and breeding decisions, cattle genomics allows producers to optimize profitability and yields of livestock herds.

Rex Animal Health is one animal genomics company that is on a mission to bring animal health to the 21st century by making veterinary medicine and husbandry data driven. Rex believes that data has the power to transform the animal health industry as it has transformed the fields of finance, weather, and human health. By bringing together data from various sources and time periods, the insights we produce can reshape: The way pharmaceutical companies discover new gene targets and develop drugs, the way veterinarians manage chronic diseases in animals and the way community understands and perceives animal health.

Advantages and Disadvantages of Livestock Technology

Sensor and data technologies have huge benefits for the current livestock industry. It can improve the productivity and welfare of livestock by detecting sick animals and intelligently recognizing room for improvement. Computer vision allows us to have all sorts of unbiased data that will get summarized into meaningful, actionable insights. Data-driven decision-making leads to better, more efficient, and timely decisions that will advance the productivity of livestock herds.

However, there are some unintended consequences of this technology. In the digitalization of industries, agriculture is often at the bottom of all charts for technology adoption. The cyclical nature of economics in the livestock industry makes it difficult for producers to justify the initial steep upfront costs of implementing these technologies.

Another challenge of the livestock industry particularly as more and more technology is being developed, is that in a dairy herd manager's office, there are often multiple computers and screens each dedicated to a different technology or records keeping program. A major need in the livestock industry is for more integration of these technologies so that there can be one platform that brings together all of this data. There are companies that are working on this and pulling data from multiple sources into one app that summarizes the data nicely so that it can be used to make well-informed decisions. Data integration and facilitating decision-making is true not only for the dairy and livestock management industry but also more broadly for the entire agriculture industry. Having data is not enough - the valuable insights should lead to actionable decisions.

What is the future of Livestock Management?

With a domestic market value of over $30 billion annually and 9 million dairy cows just in the US, investors are starting to take a major interest. Tyler Bramble, Portfolio Growth Lead at Cainthus, envisions a cow-centric approach for the future: "Today, we manage livestock for the most part around the human schedule. This technology will allow us to manage livestock from a cow-centric approach. Animals will be able to act and go about its day in a more natural behaviour and environment that it can today." Regardless of all the current challenges, the future is bright for the connected cow.

Healthcare

Technology has aided informatics in health care, data is easily collated and analysed with the aid of devices. Digitization of Health Records

Dusty and bulky paper files are slowly giving way to streamlined digital records that are securely uploaded to the cloud and made accessible remotely to patients and healthcare professionals. By this, storing, management, and transmission of data becomes easy and quick. Support for clinical decisions is made available to professionals and patients; making it possible to take better, more informed medical decisions. Digitization of health records also facilitates efficiency and delivery of healthcare to remote or inaccessible locations. This digitization has the potential to streamline processes, improve patient outcomes, and reduce costs.

Mobile App Technology in the Medical Field

Not only do patients have access to quick and accurate medical information using their handheld devices, but they can also use apps to keep track of doctor's appointments, get reminders to take their medications. Health and fitness apps help people get healthier by tracking their food intake and activity levels and offering customized solutions.

These apps can also help physicians in high-stress jobs by reducing time spent in the filing, record maintenance, and other routine tasks. Mobile apps provide access to drug information to help prevent side effects and interactions, solve problems, and improve diagnosis. Doctors can communicate directly with their patients, record their vital signs accurately, maintain logs about visits and consultations, and achieve greater procedural efficiency.

Electronic Medical Records

Electronic medical records or Electronic Health Records (EHRs) consist of digital summaries of a patient's medical records. They could include diagnoses, lab reports, and details about hospital stays, surgical interventions, and prescriptions. They offer an overview of a patient's health; permitting a more accurate diagnosis and improved patient care.

These electronic records facilitate the easy sharing of information and collaboration between labs and specialists without the time and resource expenditure on physical transmission. EHRs provide healthcare professionals with information about patient allergies and intolerances and anything else that may be relevant; particularly important if the patient is unconscious.

When properly maintained and implemented, EHR protocols can also help increase accountability and reduce malpractice.

Electronic records are easier and less time to consume to create and maintain. They make life easier for medical accountants and reduce the chance of mistakes being made.

Big Data and the Cloud

Big data is a buzzword in different industries, including healthcare. This is because the generation and collection of huge amounts of data from a number of different sources in the healthcare field are now possible. This data is then used for analytics, making predictions about possible epidemics and ultimately preventing deaths.

Cloud storage of data helps improve efficiency and accessibility while reducing wastage. This also helps in research and development of new treatment protocols and lifesaving pharmaceutical formulations. In fact, cloud services can be invaluable for medical research, providing vast amounts of research and analysis and facilitating efficient health information exchange. The cloud provides secure and cost-effective storage solutions, with backup and recovery features, but without the hassle and cost of maintaining additional server hardware.

Improved Patient Care

Technology has placed at the disposal of the healthcare community various potent tools to improve patient care. Since EHRs are easily available to physicians, they can access complete medical histories of patients and make the most well-considered medical decisions. Doctors can quickly identify possible medication errors. They can do this by using apps such as barcode scanners and patient safety improves as a result.

RFID (radio-frequency identification) technology also helps enhance patient care by providing information about the patient's vital signs, and temperature. It facilitates real-time tracking of location, communication, and identification. There are other ways in which technology has to improve patient care: 3D printing is now used to create prosthetics, customized dental devices, and even hearing aids.

Virtual reality devices and apps help ease symptoms of depression and anxiety in older people and patients with mental illnesses and can also help people with their postoperative recovery process. The augmented reality now helps surgeons and their teams perform complex operations. Virtual and augmented reality devices can also help improve health and fitness outcomes among patients. With the availability of procedures like robotic knee replacement and the use of gene therapy in cancer treatment, it is evident that the role of technology in healthcare is bound to grow exponentially in the time to come.

Information and Communication Technology

Developments in information and communication technology are constantly improving and streamlining how the healthcare industry functions and how patients interact with their care providers. Medical experts have access to comprehensive research studies as well as diverse population groups that offer new insight into genetics, diseases and care outcomes.

Care providers can compare patient data with many other patients, identify risk factors, and suggest preventive treatments using this new technology. This technology has given a huge boost to medical tourism; making it possible for patients to get in touch with specialists in practically any part of the world for consultations and second opinions; regardless of where they themselves are located. Following remote communications such as video conferences,

patients can then travel to another country to receive specialized treatment or highly advanced surgery.

Telemedicine/Telehealth

Telemedicine takes the digitization of healthcare to its next logical step; letting patients consult with specialist practically anywhere in the world. This is technology applied to the healthcare system to overcome distance barriers and facilitate critical care in emergency situations; potentially saving lives.

Telemedicine has made it possible for patients to use telemedical devices to receive home care and support using various applications and video telephony. In telemedicine, the store-and-forward feature helps transmit biosignals, medical images, and other data to a specialist to facilitate asynchronous consultations (which don't require both parties to remain present or online at the same time). This can significantly reduce waiting time for patients; speeding up treatment delivery processes.

Telemedicine facilitates remote patient monitoring by healthcare professions. This can help manage chronic conditions effectively and economically with the help of various apps and devices. Additionally, real-time interactive services make it possible for patients to consult electronically with healthcare providers. This is essentially video conferencing that helps with diagnosis, management, counselling, and patient monitoring.

FUTURE OF HEALTH CARE

The future of healthcare is shaping up in front of our very eyes with advances in digital healthcare technologies, such as artificial intelligence, VR/AR, 3D-printing, robotics or nanotechnology. We have to familiarize with the latest developments in order to be able to control technology and not the other way around. The future of healthcare lies in working hand-in-hand with technology and healthcare workers have to embrace emerging healthcare technologies in order to stay relevant in the coming years.

Be bold, curious and informed!

Are you afraid that robots will take over the jobs of nurses, doctors and other healthcare professionals? Are you scared that artificial intelligence will control the world within a couple of years? Do you have nightmares about virtual reality addicted kids and adults running around in their non-existent dream world? Are you frightened to have a genetic test because it might reveal the day of your death?

These are all half-truths, fake news and other imaginary dystopias. In a more fashionable way: alternative facts about the future of medicine. However, these all have one thing in common: the fear about the unknown place called future and what it might bring upon us.

But no matter how scary the future might seem at the moment, we cannot stop technological development; and sooner or later we will find out that whole areas of our lives have been transformed through various digital technologies. Thus, our task at the moment is to face our fears about the future with courage; to turn to technologies with an open mind and to prepare for the changing world with as much knowledge as possible.

Technology and humans hand-in-hand for a better healthcare

I honestly believe that this is the only way forward. Technology can only aid and improve our lives if we stand on its shoulder and if we are always (at least) two steps ahead of it. But if we adhere to this rule, the cooperation between people and technology could result in amazing achievements.

In medicine and healthcare, digital technology could help transform unsustainable healthcare systems into sustainable ones, equalize the relationship between medical professionals and patients, provide cheaper, faster and more effective solutions for diseases – technologies could win the battle for us against cancer, AIDS or Ebola – and could simply lead to healthier individuals living in healthier communities.

But as the saying goes, one has to be a master of his own house, so it is worth starting "the future" with the betterment of our own health through digital technologies, as well as changing our own attitude towards the concept of health as such and towards medicine and healthcare.

And what does it all look like in practice? To serve as an introduction, this article will explore 10 ways in which medical technology is reshaping healthcare.

1. **Artificial intelligence**

I believe that artificial intelligence has the potential to redesign healthcare completely. AI algorithms are able to mine medical records, design treatment plans or create drugs way faster than any current actor on the healthcare palette including any medical professional.

Atomwise uses supercomputers that root out therapies from a database of molecular structures. In 2015, the start-up launched a virtual search for safe, existing medicines that could be redesigned to treat the Ebola virus. They found two drugs predicted by the company's AI technology which may significantly reduce Ebola infectivity.

More recently, Google's DeepMind created an A.I. for breast cancer analysis. The algorithm outperformed all human radiologists on pre-selected data sets to identify breast cancer, on average by 11.5%!

These are only two of the many examples of companies using A.I. to advance healthcare from designing new drugs to disrupting medical imaging to mining medical records.

2. **Virtual reality**

Virtual reality (VR) is changing the lives of patients and physicians alike. In the future, you might watch operations as if you wielded the scalpel or you could travel to Iceland or home while you are lying on a hospital bed.

VR is being used to train future surgeons and for actual surgeons to practice operations. Such software programmes are developed and provided by companies like Osso VR and ImmersiveTouch and are in active use with

promising results. A recent Harvard Business Review study showed that VR-trained surgeons had a 230% boost in their overall performance compared to their traditionally-trained counterparts. The former were also faster and more accurate in performing surgical procedures.

The technology is also benefiting patients and has been proven to be effective in pain management. Women are being equipped with VR headsets to visualize soothing landscapes so as to help them get through labour pain. Patients suffering from gastrointestinal, cardiac, neurological and post-surgical pain have shown a decline in their pain levels when using VR to distract them from painful stimuli. A 2019 pilot study even showed that patients undergoing surgery lessened their pain and anxiety and improved their overall hospital experience.

3. **Augmented reality**

Augmented reality differs from VR in two respects: users do not lose touch with reality and it puts information into eyesight as fast as possible. These distinctive features enable AR to become a driving force in the future of medicine; both on the healthcare providers' and the receivers' side.

In case of medical professionals, it might help medical students prepare better for real-life operations, as well as enables surgeons to enhance their capabilities. This is already the case at Case Western Reserve University where students are using the Microsoft HoloLens to study anatomy via the HoloAnatomy app. Using this method, medical students have access to detailed and accurate, albeit virtual, depictions of the human anatomy to study the subject without the need of real bodies.

Another promising company, Magic Leap, will also bring its slightly different, mixed reality

headset to healthcare. Magic Leap has partnered with SyncThink for brain health, with XRHealth for developing a therapeutic platform and with German healthcare technology company Brainlab to bring its spatial computing technology to healthcare. However, no commercial products are yet available from these partnerships but we're bound to see them populate the healthcare market in the near future.

4. **Healthcare trackers, wearables, and sensors**

As the future of medicine and healthcare is closely connected to the empowerment of patients as well as individuals taking care of their own health through technologies, I cannot leave out health trackers, wearables and sensors from my selection. They are great devices to get to know more about ourselves and retake control over our own lives.

I personally use the Fitbit Ionic to monitor my sleep and track my workout. I supplement it with the Polar H10 to fine-tune my workout routines with my trainer so as to find the best exercises for my abilities. For meditation, the Muse headband helped me a lot to find the main things that I personally need for a successful meditation session. Now I don't have to use the device to try to reach mindfulness!

No matter whether you would like to manage your weight, your stress level, your cognitive capabilities better or you would like to reach an overall fit and energetic state, there is a device for all of these needs and more! The beauty of these new tech-fuelled devices is that they really make patients the point-of-care. With the ability to

"Anything's possible if you've got enough nerve." –J.K. Rowling

monitor one's health at home and share the results remotely with their physician, these devices empower people to take control of their health and make more informed decisions.

5. **Medical tricorder**

When it comes to gadgets and instant solutions, there is the great dream of every healthcare professional: to have one almighty and omnipotent device, with which you can diagnose and analyze every disease. It even materialized – although only on screen – as the medical tricorder in Star Trek. When Dr McCoy grabbed his tricorder and scanned a patient, the portable, hand–held device immediately listed vital signs, other parameters, and a diagnosis. It was the Swiss Army knife for physicians.

With the exponential progress in healthcare technology, we now live in a world where similar devices, which were once a figment of sci-fi enthusiasts, are available! The Viatom CheckMe

"If you want something new, you have to stop doing something old." – Peter F. Drucker

Pro is one such palm-sized gadget which can measure ECG, heart rate, oxygen saturation, temperature, blood pressure and more! There are also other companies working on similar devices like the MedWand which on top of measuring multiple vital parameters, packs a camera for telemedical purposes. Then there's the FDA-cleared BioSticker from BioIntelliSense which, despite being tiny and thin, can measure a wide range of parameters like respiratory rate, heart rate, skin temperature, body position, activity levels, sleep status, gait and more.

Although the currently available products are a bit far from the tricorder, we will get there soon. You will see high–power microscopes with smartphones, for example, analyzing swab samples and photos of skin lesions. Sensors could pick up abnormalities in DNA, or detect antibodies and specific proteins. An electronic nose, an ultrasonic probe, or almost anything we have now could be yoked to a smartphone and augment its features. And we have to get ready for it!

6. **Genome sequencing**

The whole Human Genome Project cost approximately $2.7 billion for the US government, which is an insanely huge amount of money. Especially if you consider that in January, 2017, DNA sequencing giant Illumina unveiled a new machine that the company says is "expected one day" to order up your whole genome for less than $100. Last year, the company's CEO reiterated that Illumina is still working towards that benchmark. This would mean that you might have a cheaper genetic test than a general blood test (for which prices vary between approximately $10-150). Mind-blowing!

Such a test has so much potential! You can get to know valuable information about your drug sensitivity, multifactorial or monogenic medical conditions and even your family history. Moreover, there are already various fields leveraging the advantages of genome sequencing, such as nutrigenomics, the cross-field of nutrition, dietetics and genomics. Some companies such as the California-based start-up, Habit, are offering personalized diets based on genetic codes.

I also took the Atlas Biomed's genetic test which proved to be quite insightful. Its analyzes, despite some being difficult to understand, provided practical calls to action. It showed me that I should have a higher intake of vitamins A and E and iron, and that I don't have any lactose, gluten, or alcohol intolerance. In addition, it also revealed conditions to which I am at risk, which is informative so as to take preventive actions.

7. **Revolutionizing drug development**

Currently, the process of developing new drugs is too long and too expensive. However, there are ways to improve drug development with methods ranging from artificial intelligence to in silico trials. Such new technologies and approaches already are and will be dominating the pharmaceutical landscape in the years to come.

Companies like Turbine, Recursion Pharmaceuticals and Deep Genomics are leveraging the power of A.I. to develop new drug candidates and novel therapeutic solutions in record time and speed up the time to market, all while saving costs and lives.

Another promising healthcare technology is in silico drug trials. These are individualized computer simulations used in the development or regulatory evaluation of a medical product, device or intervention. While the current technology and biological understanding don't allow for completely simulated clinical trials, there is significant progress in this field with organs-on-a-chip, which are already being put in use.

HumMod, or the "most complete, mathematical model of human physiology ever created", is being employed in several research projects. Virtual models have also been created by the Virtual Physiological Human (VPH) Institute which are used to study heart diseases and osteoporosis.

Imagine if we could test thousands of new potential drugs on billions of virtual patient models in minutes? We might reach this stage in the near future.

8. **Nanotechnology**

We are living at the dawn of the nanomedicine age. I believe that nanoparticles and nanodevices will soon operate as precise drug delivery systems, cancer treatment tools or tiny surgeons.

As far back as 2014, researchers from the Max Planck Institute designed scallop-like microbots designed to literally swim through your bodily fluids. Small, smart pills like the PillCam are already in use for colon exams in a noninvasive, patient-friendly way. In late 2018, MIT researchers created an electronic pill that can be controlled wirelessly and relay diagnostic information or release drugs in response to smartphone commands.

Nanotechnology is also making progress in the form of smart patches. At CES 2020, France-based company Grapheal demonstrated its smart patch that allows continuous monitoring of wounds and its graphene core can even stimulate wound healing.

As the technology evolves, we will see more practical examples of nanotechnology in medicine. Future PillCams could even take biopsy samples for further analysis while remote-controlled capsules could make the prospect of nano-surgeons a reality.

9. **Robotics**

One of the most exciting and fastest growing fields of healthcare is robotics; developments range from robot companions through surgical robots until pharmabotics, disinfectant robots or exoskeletons.

2019 was a great year for exoskeletons. It saw Europe's first exoskeleton-aided surgery and a tetraplegic man capable of controlling an exoskeleton with his brain! There are loads of other applications for these sci-fi suits from aiding nurses through lift elderly patients to helping patients with spinal cord injury.

Robot companions also have their place in healthcare to help alleviate loneliness, treat mental health issues or even help children with chronic illness. The Jibo, Pepper, Paro and Buddy robots are all existing examples. Some of them even have touch sensors, cameras and microphones for their owners to interact with them. For instance, ikki from an Australian start-up is helping children with chronic illnesses monitor their medications, temperature and breathing rate while keeping them company with music and stories.

10. 3D-printing

3D-printing can bring wonders in all aspects of healthcare. We can now print bio tissues, artificial limbs, pills, blood vessels and the list goes on and will likely keep on doing so.

In November 2019, researchers at the Rensselaer Polytechnic Institute in Troy, New York, developed a method to 3D-print living skin along with blood vessels. This development proves crucial for skin grafts for burn victims. Also, helping patients in need are NGOs like Refugee Open Ware and Not Impossible which 3D-print prosthetics for refugees from war-torn areas.

The pharmaceutical industry is also benefiting from this technology. FDA-approved 3D-printed drugs have been a reality since 2015 and researchers are now working on 3D-printing "polypills". These contain several layers of drugs so as to help patients adhere to their therapeutic plan.

Food for thought

We are truly living in revolutionary times for healthcare thanks to the advent of digital health. Our mission is to spread the knowledge and developments in healthcare which will usher the real era of the art of medicine. Join us in this mission by sharing our articles and your thoughts with us!

Manufacturing

The shift to smart factories, manufacturing innovation trends,

Today, innovation has been permeating all facets of our life. Manufacturing is one of the most vivid examples of the inevitable power of innovations and industrial breakthrough. We are already on the verge of Industry 4.0 driven forward by innovations in automation, AI, and IoT. No doubt, the way factories operate is constantly changing, leaving significant footprint on the world around us.

What does Industry 4.0 stand for?

Industry 4.0 presupposes integration of manufacturing automation and data exchange to encourage innovations and smart factory. Such factories will be controlled by a virtual production line that runs systems and monitors and completes their physical processes. Communicating in real time and ensuring the quality of its operations, connected software systems will be running physical manufacturing automatically. So, it's quite logic that smaller, more agile companies are the ones rapidly innovating in the world of manufacturing.

The cornerstones of manufacturing innovation

Innovations strategy and culture serve for encouraging innovations in a thoroughly tailored systematic way. The creation of a culture of innovation helps the cause of serial innovation. To successfully create and commercialise inventions, companies must commit themselves to the process of innovation. Favourable atmosphere for innovations is of crucial importance, for its exactly what makes creativity to prosper, boosting innovations therefore.

"Anything's possible if you've got enough nerve." –J.K. Rowling

Design is the first outside manifestation of an innovation that gives life to an innovation in future. Without a design, an innovation remains in the mind of the innovator. Value delivery through innovation. Engineering makes production on a large scale possible, allowing it to become transferable to many locations, letting the finished product reach the masses. Engineering and production are the most important means to deliver on the promise.

So, what are 7 manufacturing innovation trends that are powering the shift to smarter factories?

1. **Virtual manufacturing**

In any manufacturing process, being able to do something exactly right the first time is ideal. Knowing precisely what an outcome will be, based on the decisions, can eliminate any wasted time or resources, by being able to essentially manufacture something virtually. That is, a team can perform its manufacturing process digitally,

"If you want something new, you have to stop doing something old." – Peter F. Drucker

with each step of the process done visually in a near-animated process.

They can understand how the procedure will go before the actual physical manufacturing takes place. This process is already under way in many industries, including auto manufacturing. It's allowing companies to save money, simplify operations and get products to market much faster than ever before.

For instance, Boeing, has already started using augmented reality in their assembly processes. They've been using Google Glass and Skylight software to give technicians valuable insight and instructions when completing complicated wiring tasks.

It's likely that other industries will adopt similar uses for the technology, allowing their output to be built with the utmost precision.

2. **Micro manufacturing and machine vision error detection**

Micro manufacturing is used to create the tiny components in a variety of different devices, including cutting-edge medical equipment that allows doctors to treat patients without resorting to invasive surgery. With technology as mobile and lightweight as it is, the micro molding process is the essential method for creating the internal machinery that designers are dependent on to create these products.

On the other hand, new technologies are allowing us to detect faults in production processes. To illustrate, Landing.AI is a company that creates smart technology that can find the tiniest of faults in circuitry of a machine that might not be immediately apparent to a human. As machinery in smart factories will all be connected and communicating with each other, the AI will send an alert when a fault is found, immediately halting the machine in question so that it can be fixed.

3. **Industrial robotics**

Today, industrial robots are very sophisticated with the ability to be easily programmed to handle more than just a single, repetitive job.

Thus, in 2016 General Motors announced that they would be adopting a robotic glove designed by NASA, for use by their factory workers. The glove is based on a design that was created for use on the International Space Station. There are numerous benefits of such a glove for the workers, including strain reduction for the workers' arms and better grip when lifting heavy items.

While robot workers are becoming increasingly common, human workers are likely to wear robotic implements like these gloves to aid in their work in future.

4. **Sensors on the workplace**

With factories gradually filling with autonomous machines, it's imperative that there are proper structures in place to prevent any accidents. One key component of this is sensors which allow machine-to-machine communication.

For example, TE Connectivity is a company producing sophisticated sensors that transmit data between industrial machines and smart devices, keeping everything running smoothly.

5. **Fully automated warehouses and smart recycling**

The Ocado Warehouse is a perfect example of the warehouses of the future. The automated warehouse is staffed by a team of robots that empty, transport, and replace batches of products. Once the robots have chosen and transported products to a picking station, the products are then prepared for shipping by other robots and human workers.

Another example is Apple's recycling robot Daisy that can strip apart iPhones and allow the materials to be reused. Such types of robots allows customers to safely dispose of their discarded smart devices. Secondly, it reduces the resources spent on sourcing the materials for new devices, as the materials can be taken from older models and reused.

6. **3D printing**

The "printing" is manufacturing of a tangible object from an alloy or plastic based on a 3D image imported into the printing machine. This process is changing the importance of funding in manufacturing and in how long it takes to bring a product to market.

Both businesses and consumers will be benefiting from these manufacturing innovation trends, as people's health and longevity are improved, communication becomes easier and manufacturing is simplified for a variety of companies.

MX3D's six-axis robot arm is one of the most exciting examples of 3D printing. Its arm can print and construct complicated metalwork in mid-air, from basically any angle. This improves not only the speed at which the parts or object can be produced, but also the structural integrity.

7. Exoskeletons for workers to ensure safety at work

Full-body exoskeletons are another piece of wearable tech designed to protect workers and increase their strength. It's the perfect compromise between an all-robot staff and protecting human jobs, as it gives human workers advantages in strength and stability usually reserved for their mechanical counterparts.

The Ekso vest is just one example of an exoskeleton designed for factory workers. The vest is already in use in a number of Ford factories, and employees have praised the device for allowing them to conserve physical energy throughout the workday while allowing them to lift weights that they otherwise would be unable to manage.

"Anything's possible if you've got enough nerve." –J.K. Rowling

We live in the era of constant transformations. Ability to be flexible and agile is one of the key qualities for success. Product and process innovation have been altering manufacturing already for decades. It goes without saying that the future will witness even more breath-taking technological disruptions, as research around nanotechnology and analytics start to impact numerous manufacturing applications.

Companies with an innovation edge featuring a clearly developed forward-thinking approach and readiness to rapidly adapt to innovative technology will have a strong competitive advantage. Only strong players will get a chance to fulfil established goals, provide goods and services for untapped markets, and, finally, to stay ahead of the competition!

"If you want something new, you have to stop doing something old." – Peter F. Drucker

Power

Would you buy a home without functioning power outlets? We've come to depend on electric power for many of our day-to-day activities. When there are outages, we feel the impact of that dependency, followed by a sense of relief when the power's back on again.

Electricity has helped us stay healthier, work more efficiently and live life around the clock. Because electricity has had such a positive influence on our lives, science and industry researchers are constantly finding ways to provide electric power more easily and inexpensively. As a result, innovations in electric power have made the industry cleaner and more efficient throughout its history, and made electric service available to millions of homes.

WIND TURBINE

For millennia, people have harnessed the power of the wind to accomplish tasks. For example, merchants once relied on the wind to sail the world. Also, old windmills, once used to mill cereals, are an iconic part of Holland's landscape. With our lives cantered around electricity, modern scientists have found innovative ways to convert the kinetic energy from the wind into electric power.

Today, around the world, the wind-electric turbine is becoming as iconic as the Dutch windmill. A wind turbine typically consists of a large, three-bladed propeller, called a rotor, atop a tower that's high enough that nothing blocks it from the wind. The turbine has a drive train like a car's engine that includes an electric generator. The electricity generated gets added to the electric grid, which powers hundreds of homes and businesses in a geographic location.

"Anything's possible if you've got enough nerve." –J.K. Rowling

One small wind turbine can power a single home or small business. These smaller versions have rotors between 8 and 25 feet (2.4 and 7.6 meters) in diameter and can stand up to 30 feet (9.1 meters) in the air. Wind farms are becoming increasingly common in large open spaces. You can see some of these farms during a drive or flight through the Western United States, with thousands of giant white wind turbines stretching across hillsides as far as the eye can see.

HYDOELECTRIC DAM

Hydroelectric dams are the oldest technological innovation in our countdown. During the early 1900s, 40 percent of the electricity used in the United States came from hydroelectric dams. Today, hydropower accounts for nearly a quarter of all electricity used worldwide. In addition, the physical structures themselves are marvels of human engineering and construction, drawing photographers and tourists, according to the U.S. Bureau of Reclamation.

"If you want something new, you have to stop doing something old." – Peter F. Drucker

Hydroelectric dams work by holding back massive amounts of water and allowing a limited amount to flow through the dam. The water pressure created by limiting this flow is tremendous, and hydroelectric plants harness this pressure to turn turbines attached to electric generators. As with wind turbines, the electricity generated from a hydroelectric dam is added to the electric grid associated with the dam's geographic location.

A hydroelectric dam supplies the electric grid with several hundred kilowatts to several thousand megawatts of electricity per second.

Despite its age, the future is bright for hydropower as the hydroelectric dam gets a 21st-century makeover. Researchers are finding ways to improve the efficiency and environmental impact of hydropower by improving existing dams and building new dams.

SOLAR CELLS

While wind and water can be used to generate power through movement, the sun provides a significant amount of energy in the forms of heat and light. Solar cell technology, called photovoltaic (PV) cells, convert that light into electricity. These PV cells contain semiconductor materials such as silicon. Electrons in the semiconductor move when the material absorbs the light.

Unlike the water and wind power technologies we've covered, solar cells are versatile in size and portability. Large solar panels with hundreds of cells can be built in a factory then sold to stretch out across land or mount on a rooftop. These large panels are used to power homes and businesses and must be replaced after about 30 years. Small solar panels with only a few cells gather enough energy to power standalone devices, like calculators and outdoor lighting.

Despite being a clean, renewable energy source, sunlight alone is not sufficient for those who want to use electricity at night or on cloudy days. In most cases, solar panels are a supplemental power source for a building that's already attached to the electric grid. A few people, though, choose to go "off the grid" entirely and use rechargeable batteries to store solar-generated electricity when the sun's not shining.

So far, we've looked at innovations that make the most of renewable energy sources. Next, we'll look at an innovation making use of the most efficient non-renewable source of energy known today.

2: **Nuclear Reactors**

Nuclear fission is the process of breaking apart an atom, releasing the energy that holds the atom together. In the 1950s, nuclear fission of the radioactive isotope uranium-235 made energy cheaper and more efficient to produce. A nuclear reactor is a structure that produces this fission process from uranium-235. Nuclear power plants include one or more reactors along with large and complex mechanisms for cooling and containment.

The nuclear reactor itself is the key innovation here. The reactor controls the fission process from a very small amount of uranium-235 and channels the energy to heat rods which, in turn, heat water to produce steam. The steam moves a turbine and turns an electric generator, similar to the way wind and water turbines work. So, in essence, a nuclear plant is just a steam plant powered by its nuclear.

By using nuclear power, the world uses less of other resources, like coal and oil, to heat the water and produce steam. Despite this advantage, concerns still plague the minds of sceptics. Concerns include the safety of people who live and work in and around nuclear plants and the potential hazards of nuclear waste disposal. In addition, several notorious nuclear reactor disasters around the world have tainted the reputation of this energy source.

None of these great innovations in electric power would be widely available without the top innovation in our list. Let's check that out now.

1: **Electric Grids**

Topping our list of innovations is the grid itself. When people say "the grid," they're referring to a network of electric power resources covering a certain geographic area. Some grids are attached to other grids to share resources in case of emergency. Most people who use an electric power source attach to an existing grid through power lines.

A grid is a massive electric infrastructure consisting of power lines, power stations, substations and transformers. Grids in the United States are controlled by a combination of public and private entities. The public entities are the state and federal bureaucracies that enforce laws regulating the industry. The private entities are utility companies that provide access to a grid and meter the power used by each home or business. These controlling forces determine the price a user must pay for each kilowatt hour of electricity on that grid.

Grid technology continues to expand even as people look for other energy sources to fill it. For example, smart grid technology is under development to improve efficiency in how power is monitored and metered for each customer. Also, grid storage at stations and transformers can also keep energy in reserve to help prevent blackouts during some of the grid's normal operational hiccups.

NANOGENERATORS: USING THE POWER WITHIN US

While wind and water movement can power many homes, nanogenerators use your body's movement to produce electricity on a smaller scale. Nanogenerators are tiny devices with a piezoelectric material, meaning the material creates an electric current just by bending or stretching it. This power could come from the subtle of movements, like a heartbeat or lungs expanding and contracting. Eventually, nanogenerators might be used to power pacemakers, eliminating the need for repeated surgeries when batteries fail. Soon, look for nanogenerators available in clothing, allowing you to power your portable music player just by breathing and moving.

Banking and Finance

Financial Institution have been reluctant to update their systems – and for good reason. The current systems that they use are the product of years of continued innovation to meet immediate customer requirements. But this has resulted in siloed systems being used for the transaction, savings, investment, and loan accounts. This is not suited for the digital age when the competition for banks is coming from technology based FinTech start-ups.

Banks and other traditional financial service provider have had to respond with an array of digitization and innovation initiatives. These initiatives employ cutting-edge technologies to ensure a customer-centric perspective rather than the traditional focus on products, real-time intelligent data integration rather than slow analysis being performed after-the-fact and open platform foundation. The new direction for the future is as follows:

1. Augmented Reality
2. Blockchain
3. Robotic Process Automation
4. Quantum Computing
5. Artificial Intelligence
6. API Platforms
7. Prescriptive Security
8. Hybrid Cloud
9. Instant Payments
10. Smart Machines

Augmented Reality

Immersive technologies such as Augmented, virtual, and mixed reality are enhancing customer experience across the board. So why can't they do the same for banking customers?

The possibilities of the implementation of augmented reality technology in banking sector are only limited by imagination, though these are still in an early stage of development. The end-state is to give customers complete autonomy in actions and transactions they could perform at home. Hybrid branches are envisioned by technology experts who believe that bank branches as we know them today are a thing of past.

One of the implementations of augmented reality technology in banking sector, that is already live, has been made by the Commonwealth Bank of

Australia. They have created a rich date augmented reality application for their customers who were looking to buy or sell a home. It provides them with information like current listings, recent sales, and price tendencies to help the customer make better decisions.

The National Bank of Oman has its augmented reality app that enables customers to locate the nearest branch or ATM. Deals and offers can also be uncovered as bank customers walk the streets of Oman and use the camera of their smartphone to bring together their real-life surrounding and an AR projection.

2. **Blockchain**

Blockchain is a catchall phrase used to describe distributed ledger technologies. You could think of it as a distributed database with no DBA involved. It allows multiple parties to access the same data simultaneously, and at the same time ensures the integrity and immutability of the records entered in the database. At present, leading banks around the world are exploring proof of concept projects across various aspects of banking and financial services. The first major implementation that we are likely to see is in the areas for clearing and settlement.

Another major area in which banks will see a huge saving by using blockchain technology is KYC (Know Your Customer) operations. Business models being developed now would turn KYC from a cost centre into a profit centre for banks – as they would come to rely on a shared blockchain for this activity. Syndicated loans, trade finance and payments are other areas where the smart contracts on blockchain could be highly effective.

3. **Robotic Process Automation**

The volume of unstructured data that the bank has to process is increasing exponentially with the rise of the digital economy. This is not just banking transaction data, but also other behavioural data that could potentially allow the banks to improve and innovate customer experience.

This has made bankers realize that they need to find technologies that can mimic human action and judgment but at a higher speed, scale, and quality. The answer that has emerged is a combination of various technologies that enable cognitive and robotic process automation in banking.

These technologies consist of machine learning, natural language processing, chatbots, robotic process automation, and intelligent analytics in banking that allow the bots to learn and improve.

In the years to come, we would see the current cognitive capabilities being bundled with the robotic process automation to achieve even better results. This is already being implemented in point-of-sale solutions that automatically suggest marketing promotions that would be most effective for an individual customer.

4. **Quantum Computing**

Quantum computing is a way of using quantum mechanics to work out complex data operations. As is common knowledge today, computers use bits that can have two values – 1 or 0. Quantum computing uses "quantum bits" that can instead have three states – 1 or 0 or both. This unlocks exponential computing power over traditional computing – when the right algorithm is used. This represents a huge leap in computing power, but any commercial implementations are still decades away.

5. **Artificial Intelligence**

The explosive growth that the last decade has seen in the amount of structured and unstructured data available with the banks, combined with the growth of cloud computing and machine learning technologies has created a perfect storm for Artificial Intelligence to be used across the spectrum of banking and financial services landscape.

Business needs and capabilities of AI implementations have grown hand-in-hand and banks are looking at Artificial Intelligence as a differentiator to beat down the emerging competition. Artificial Intelligence allows banks to use the large histories of data that they capture to make much better decisions across various functions including back-office operations, customer experience, marketing, product delivery risk management, and compliance.

Artificial intelligence would revolutionize banks by shifting the focus from the scale of assets to scale of data. The banks would now aim to deliver tailored experiences to their customers rather than build mass products for large markets.

Instead of retaining customers through high switching costs, banks would now be able to become more customer-focused and retain them by providing high retention benefits. Most importantly, banks would no more just depend on human ingenuity for improving their services. Instead, performance would be a product of the interplay between technology and talent.

6. API Platforms

The time when banks could control the whole customer experience through a monolithic system that controlled everything from keeping records to every customer interaction is long gone. Both the regulatory requirements and the revolving customer needs have turned this humongous system into dinosaurs.

Today banks need to instead build "banking stacks" that allow them to be a platform to which customers and third-party service providers can connect to deliver a flexible and personalized experience to the end user. To do so, they can use API platforms for banking.

API Banking Platform is designed to work through APIs that sit between the banks' backend execution and front-end experiences provided by either the bank itself or third party partners.

This allows the banks to adopt completely new business models and use cases (for example, enabling salary advances) and experiment with new technologies like blockchain at low cost. APIs also help banks to future-proof their systems as the front-end is no more tightly coupled with the backend.

7. **Prescriptive Security**

The nature of cyber risk changes at a great speed. This makes the traditional approaches to risk management obsolete. It is now clear that it is impossible for organizations to eliminate all possible sources of cyber threats and limiting the attack footprint at the earliest is the best way to deal with these. The banks will have to be nimble in the way they approach cybersecurity.

"Anything's possible if you've got enough nerve." –J.K. Rowling

Increasingly banks are deploying advanced analytic, real-time monitoring and AI to detect threats and stop them from disrupting the systems. The use of big data analysis techniques to get an earlier visibility of threats and acting to stop them before they happen is called prescriptive security.

While the disruption brought by implementing the new technique may lead to an increase in vulnerability at the start, this is the way forward to stop the ever increasing data breaches that various organizations are reporting.

8. **Hybrid Cloud**

One of the biggest challenges that the digital age has brought to banking is the need to respond quickly. The constantly evolving market that banks operate in requires them to be as agile as possible. They need to be able to provide resources across the enterprise in a timely manner to address business problems faster.

"If you want something new, you have to stop doing something old." – Peter F. Drucker

High performing banks have discovered that the most cost-effective way of achieving this is through an enterprise-wide hybrid cloud. This allows them to pick benefits of both public and private while addressing issues like data security, governance, and compliance along with the ability to mobilize large resources in a matter of minutes.

Hybrid cloud also allows banks to offer innovative new offerings to its customers. For example, ICICI Bank has partnered with Zoho to allow businesses to automate the basic reconciliation process through Zoho Books, a cloud accounting software. The partnership does away with the need for data entry and makes it easier to offer multiple payment options to the customers.

9. **Instant Payments**

As the world moves towards a less-cash economy, the customer expectations around payments have changed dramatically. Both customers and business expect payments to happen instantaneously, and this is where instant payment systems step in.

Instantaneous payment is a must if online payments need to replace cash transactions. Therefore, banks around the world are finding ways of providing their customers options for instant payment, even when the infrastructure required for the service is lacking.

10. **Smart Machines**

You must have already seen assistants like Amazon's Alexa and Google Home in action. Can you imagine the impact these could have on banking applications?

In fact, Bank of America has already developed Erica as a virtual assistant specifically for banking operations. These smart machines are beginning to act as digital concierges for the customer in interacting with banks as well.

Banks will have to invest in digital engagement to ensure long lasting relationships with the customer. Remember that customers will gravitate towards banks that are easiest to work with when they are using technologies that they have become habituated to.

Procurement

For governments to carry out their day-to-day functions, procurement — or their ability to purchase goods and services — is critical. It is both a service function and a strategic policy tool which can help achieve a broad range of social and economic welfare objectives. It cuts across all areas of public administration and builds on cooperation among multiple public and private stakeholders.

For procurement to be more effective, then, it needs to innovate. Promoting innovation in procurement means processes that are transparent and efficient, and that facilitate equal access and open competition. Innovative solutions to public service needs are instrumental to delivering better services with long-term value for money.

Governments, as the major purchasers in every market, can also drive innovation in the private sector by stimulating their response to current and future service needs, smart regulation, and demand for innovative solutions. They can influence various industries' investments in new skills, equipment and research and development, which, in turn, support growth and competitiveness. Many companies have already realigned their own procurement to put greater focus on aspects such as time to market and product success rather than traditional aspects like savings and contract compliance.

Based on the potential of innovation, the changes in the procurement function should be centered on providing solutions to complex problems through collaboration of independent, multiple actors who work together co-create solutions that will help ensure the most effective procurement performance for the best possible outcomes.

To accomplish innovation both in public and private sector procurement, it is necessary to understand that:

Innovation should not be about products or about money. It is about people, ideas, and leadership. It's about understanding clearly the unmet needs that we want to target (be that resolving a problem, or improving value for money) by listening, collaborating, capturing the potential for enhancing outcomes, and scaling it up.

Simplicity is often the trademark of innovative solutions, but implementing them requires a deliberate strategy to use procurement to meet business objectives, clear processes for the identification of needs (stated as outcomes), timely and effective ideas and implementation management, and measuring of results.

An important and enabling element for innovation is a culture that encourages intelligent risk taking and recognizes and rewards innovation. Innovation, especially in procurement, is not just "invention" in an environment where a thousand flowers bloom. It is useful when specifically, needed, or desirable, possible with technology, and viable in the context where it is supposed to be implemented. Therefore, most organizations where innovation happens consistently rely heavily on big data and technology, value technical excellence, and focus on external customers, rather than internal clients.

Innovative procurement will also mean a better understanding of how institutions and markets behave in different contexts. It will allow us to better use the whole procurement cycle — from strategic planning to contract management — to stimulate innovation through more informed, evidence-based decision-making.

People and institutions have always been innovating. For those of us working on procurement – from international organizations, governments, private sector or civil society – we should all be agents of change, defining new ways of collaboration that allow us to unleash a true transformational way in which procurement can deliver results.

Warehouse

In case you are wondering how the Internet of Things can contribute to increasing the productivity of your store's warehouses, there are dozens of projects that are tested and launched each year. When designing an IoT-based product, think of the objectives you want to accomplish with the technology. Here are the most common goals managers achieve by implementing the Internet of Things:

1. **End-to-end inventory tracking**

From the product's arrival to the warehouse to its delivery to a shopper's doorstep, the manager will be provided with status updates, a location within IoT tracking system for warehouse, and notifications in case of complications.

2. **Vision picking**

Smart glasses help employees effortlessly pick items and transport them to a different zone in the warehouse. Since connected wearables have intuitive interfaces, learning to use them takes less time than memorizing the location of each type of goods and having to reach it manually. Vision picking systems reduce the number of paper documentation and track product lots automatically, thus increasing the productivity of the warehouse.

3. **Automated tasking**

Tasks, like picking and packing, are monotonous and tiresome — thus, the odds of human error are higher than in more demanding tasks. IoT and smart warehouse technologies help automate repetitive assignments and allocate the workforce more efficiently. By introducing IoT to the warehouse, store managers will be able to reduce order inaccuracies and inventory damage.

4. **Data analytics**

Despite the wide range of connected devices, the true value of the Internet of Things in warehouse management lies in processing and analysing collected data. Connected data analytics systems are widely used in warehouses to ensure safety. An IoT-enabled platform can, for instance, monitor the warehouse floor by gathering the information provided by sensors and alert a store manager in case of anomalies.

In the last decades, using the Internet of Things for warehouse management became more than a promising concept — companies started implementing sensors, RFID tags, device-to-device communication, and other forms of connectivity to manage daily tasks.

As of now, there is no lack of successful warehouse IoT applications in warehouse management. Here are a few that, in our opinion, are the most rewarding ways to leverage the full potential of the innovation:

RFID technology in IoT. RFID tags can store considerably larger data volumes than barcodes. That's why they provide managers with more information on every lot — its size, manufacturer, expiration date, serial number, production line, and so on. An average RFID reader has a higher speed than the one for barcodes and can scan up to 200 tags at once.

IoT-integrated management systems. IoT for warehouse tools are a cut above traditional ERP (Enterprise Resource Planning) system. Instead of gathering inventory data manually, staff members can outsource the task to a range of connected sensors or RFID tags. The data is then stored on a cloud-based platform, processed and analyzed. Finally, a user sees condensed inventory or other warehouse-related data in a clear way via a visual dashboard.

Wearables. By using connected devices, warehouse workers can instantly identify products and packages. Wearables will track the accuracy of staff product picking to assess the efficiency of corporate training and keep track of employees' individual performances. Other features include a heartbeat and vitals monitor that help ensure employees are not exhausted at work.

Sensors help warehouse managers maintain better control over the merchandise inside the warehouse and out. By integrating them into the supply chain, managers will be able to track goods on every stage of delivery and monitor temperature and humidity inside the truck. Light motion, humidity, and temperature sensors are all widely used by store managers. Adopting sensors that would track drivers' vitals is another proactive way to neutralize supply-chain-related risks.

Benefits of Implementing IoT in Warehouses

If you are wondering whether the Internet of Things is worth implementing in the warehouse, here are the reasons to consider embracing the technology as soon as possible:

Improved transparency. Supply chain transparency is not only a way to streamline internal operations, but a powerful strategy of connecting with customers as well. IoT allows store managers to collect warehouse and supply chain data in real-time and share it with customers. This way, a shopper will not be disappointed if a wanted product is unavailable — a system will also send a notification once a store is restocked.

Improved last-mile delivery. Last-mile delivery covers over 30% of all delivery costs due to the high reliance on traffic, drivers' skills, and fuel costs. With the Internet of Things, delivery trucks can collect orders more efficiently, fully using all the available space.

Predictive maintenance. A predictive maintenance system detects the early signs of equipment malfunctions, allowing store managers to prepare spare machinery and avoid downtime. The Internet of Things enables reducing downtimes and machine repair expenses, facilitating warehouse management considerably.

Real-time product tracking. IoT solutions for warehouse management provide real-time data on product locations, transportation conditions, the integrity of packaging, and so on. Thanks to instant updates, store managers can ensure no inventory is lost during transportation and ensure that supply chain vendors are managing deliveries responsibly.

Higher employee productivity. IoT platforms help staff get instant, on-demand assistance — this way, they can perform a higher number of tasks per day. Connected devices help navigate the warehouse, prioritize tasks, and identify the right packages.

Examples of IoT Applications for Smart Warehousing

When implementing IoT solutions in warehouses, consider taking pointers from industry leaders that have already launched connected applications. Here are the top cases of global IoT adoption:

1. Amazon Warehouse Automation

Amazon has recently launched a semi-automated warehouse, where robots work alongside human employees. Basic tasks, like moving packages around or scanning barcodes, are outsourced to technology. Sorting through packages and moving objects of complex shapes (bottles, for instance) is still a part of human jobs. Amazon's automated warehouse employs over 400 robots and hundreds of human employees

2. **Alibaba Warehouse Automation**

To deal with bottlenecks on 'The Singles' Day', the company's global shopping fest, Alibaba launched a fully robotic warehouse in 2018. There are over 700 guided robots designed to transport parcels across the place and deliver goods to delivery trucks. According to the president of the e-commerce giant, fully automating a warehouse helped save a ton of time and ensured faster, error-free product deliveries.

3. **Ocado Warehouse Automation**

A popular, online-only British grocery store uses connectivity to automate basic warehouse activity. The company uses simple bots to automate basic tasks — moving goods around and lifting them up. For the company, using space efficiently is a top priority.

That's why developing algorithms that can lift boxes higher than a human is capable of has been Ocado's priority. The inventory processes over 60,000 orders per week and is active 24/7.

4. **DHL Smart Warehouse**

DHL has piloted a range of innovations at its warehouses. The company uses smart glasses, robots, drones, autonomous vehicles, and so on. With constant visual support, it's easier for a worker to identify products and sort through parcels.

Recently, DHL has teamed up with Cisco to create a platform that would monitor supply chain activities in real-time — the project is still in development. Warehouse management encompasses a wide range of activities across various locations. Aside from inventory monitoring, managers have to oversee the supply chain, fill in the financial documentation, and fight the talent shortage. The Internet of Things helps business owners heighten the security of warehouses, track the location of goods, detect the warning signs of weather changes and equipment malfunctions. You will be able to streamline internal operations and update customers efficiently.

Logistics

Logistics is a complicated sector. Various negative outcomes have given rise to various innovation in the logistics industry. Transforming the e-commerce industry, innovation in the logistics industry is connecting people all over the world, through a managed supply chain. These logistics has made the e-commerce industry's efforts worth it. They have also reduced cultural difference among people. The logistics industry has witnessed a lot of twists and turns in their journey. They have pulled off some major innovations in the way they do business.

Technology has left people awestruck and has proved that there is nothing that technology can't do. These 7 logistics automation programs will transform the way you look at modern day logistics.

A Virtual calling Technology– A innovation in logistics industry

Using an Ai driven technology, logistics are introducing software with virtual calling systems. In this method, the company automates a call to its customers to confirm the specifications of a product.

An NDR team takes full responsibility for this. They take care of all the non-delivery of the product. They also resolve all the delivery related issues. This technology has shown to reduce return and refund rates to almost 15%.

Automation technology

The automation technique used by different logistics companies uses data-driven software. This is done to improve its operational efficiency. It also helps in streamlining various other logistics operations. These labelling, packaging, and sorting, along with all other logistics functioning use automation techniques. There are a lot of logistics who even use automation for their entire cargo functioning.

Because of the automation, the overall productivity of logistics increased up to almost 30%. It has added a lot of value towards revenue generation from these online businesses.

The Robotic Technology

The possibilities have been made endless, and several jobs are being performed at once. With an increase in supply chain demands, different logistics are reducing their manual labor forces and switching to robotics. The rising importance of these logistics is leading to increased adoption of robotics in different warehouses.

3D Printing Technology

Another major innovation transforming the logistics and supply chain management is 3D printing technology. It rebalances the labour cost and also reduces inventory management and transportation costs. In this process, the supply chain is localizing the outsourced supply process and is also building its own in house logistics.

The Blockchain Technology– innovation in logistics industry

The blockchain is the process of decentralizing the computer networks. This helps them to keep a permanent record of the transaction.

It is an extremely effective technology that contributes to a lot of logistic sectors. they allow cost-cutting, data verification, asset tracking and many more. they also help in maintaining accountability for the different logistics sector. They also help in taking care of smart contacts and provides compliance.

The digital logistics technology

There are many technologies innovating the logistics business, and this digital marketplace technology is one of them. This technology not only addresses the potential mismatch but also resolves all the issues between supply and demand.

This is leading to a better platform for all the e-commerce business. They are helping different companies to better utilize their assets. Not only this, but they are also improving the sales rate for retailers and reducing the return ratio.

The on-demand and crowd shipping technology

Various food outlets are using this technology for providing home delivery services. This on-demand technology leverages the bulk supply at a much faster rate and improves the last mile delivery market. The crowd shipping process involves the delivery of individual parcels during an ongoing journey that increases the supply rate in the market.

Use of Autonomous Vehicles

This is a great innovation that holds the potential to revolutionize the logistics industry. The autonomous trucks coming into the market are not only lowering the transport cost but are also ensuring the right delivery at the right time.

Use of the alternative fuels

The future predicts the use of alternate fuel. They will provide much more power to the trucks and vans in the years to come. The diesel is probably one of the best fuel and various experiments are still going on to find out a better alternative.

Wearable Technology as an innovation in logistics industry

This technology is soon going to be a must have a form of technology. The technology doubles the speed of work like packing and shipping. It reduces human labor. It also automates the entire process.

The process has increased logistics efficiency by 25%. It can be expected to improve in the coming years. The process of wearable technology not only reduces hassle but also help in increasing customer satisfaction.

The Use of Drones

This provides a promising benefit to the logistics industry. The process not only eases the coordination from shipment to delivery but are also having a great impact on effective product delivery and can reach every part of the country.

The areas include the rural, urban and even extremely remote areas.

The business is changing, along with the future of logistics. These innovations are not only becoming some of the best creations of human beings but are also proving that nothing is impossible. iThink Logistics is part of one of these innovations. The AI-driven technology used by this organization is reducing return and refund rates. It is also establishing a better customer-seller relationship.

Physical Security

Talk to physical security specialists and there's one thing they want you to understand — that stuff you see on TV where they freeze a frame of surveillance video and zoom in down to the pores, that's just not possible. Yet.

The gap between Hollywood and security technology in the real world is closing, though more slowly than some would like. In the realms of surveillance and video analytics, the security world is abuzz with advances in the capabilities of current technology. And the near future holds the promise of developments such as virtual command centres. Here is what facility managers can keep an eye on now as they consider tools to help augment their security team's capabilities.

Robots and drones

When Mark Schreiber, president and principal consultant of Safeguards Consulting, talks about security robots, he prefers to use the term "unmanned ground vehicle" to more accurately reflect the capabilities of today's technology. He currently sees a limited use case for security robots, which is tied to enhancing active surveillance efforts. Guard tours of a facility or campus can get to be monotonous work, where fatigue and boredom increase the chance for error. But a security robot could patrol the same space in a consistent manner, over and over.

Facility managers should keep any visions of Robocop well in check, however. "It will take some time to really get out the different nuggets of value," says Schreiber. "It's only the advanced high-tech organizations that are really going to start running with the ball initially."

One early adopter of security robots is Microsoft, where the robot program at the company's Redmond, Wash., headquarters has been successful enough that the technology will be spread to other Microsoft campuses. "Think about it more as a mobile sensor platform," says Mike Anderson, security program manager, Microsoft global security. "That's all it really is." Sensors mounted on the robot can be deployed to detect any number of factors, from unregistered vehicles in a parking lot to unusual heat patterns in a data center. At Microsoft, some robots were deployed with an emergency communication call button, so an individual needing immediate assistance in a parking lot could run up and open a communication line with security, for example.

"The technology is still pretty young," says Anderson. But he expects to see more robots patrolling places like malls or large corporate campuses in the next two to three years, "doing things that would enhance security operations, not replace them." Drones, or unmanned aerial systems, have a very different impact. "They're a totally different type of platform for the same type of security sensors and technologies that we use in a fixed platform in many different locations," says Schreiber. A surveillance camera on a fixed pole has a distinct value, primarily gathering forensic data. But put the camera on a drone and you now have a very fast mobile platform for active surveillance and emergency response.

On a large campus, a drone can be on site within minutes where a first responder could take much longer. "With a drone platform, we can get there in a fraction of the time, immediately get video and other sensors on that incident, and quickly evaluate what response is needed," says Schreiber. All that without having to endanger a human life, as would be the case to evaluate a dangerous situation, like an explosion.

More than deploying a drone security team, Anderson says most companies are focused on developing an anti-drone plan. How can facility managers detect a drone on their campus? How can they detect where the signal is coming from, take over the drone, and land it? Anderson says facility managers should start by establishing the drone policy for their campus, and then worry about anti-drone defences. For example, at

Microsoft, the company's own researchers on drones and robotics have to get security's permission first to operate on campus. "They just want to fly their drone wherever they want, and you can't have that," Anderson says. "We had to establish a pretty strict policy."

Video analytics is one area in physical security where the public's expectations initially far outstripped the capabilities of the technology. "Initially, when video analytics first came out, it had a lot of hype," says Schreiber. "We're at a point where the technology has advanced so much that there's a video analytics application for almost any type of large facility."

Passive video surveillance is common and gathers video for forensic purposes. Video analytics come into play with active video surveillance. Simple solutions, says Schreiber, include monitoring an area for activity when there should be none, or sensing direction of travel in trip-wire setups.

Outdoors, a simple analytics application might be detecting motion at the perimeter. Algorithms have advanced to the point where the movement of trees or rain no longer set off false alarms.

However, the current state of off–the-shelf analytics is still relatively limited, says Coleman Wolf, security practice leader at Environmental Systems Design. "It's a tool and it can certainly do some great things, but people have to understand the limitations of that tool," he says. "I always caution clients that it doesn't replace human thought process. It should be used as a sensor, but you really still need a person to be part of the assessment."

In the near future machine learning for video analytics, and physical security in general, will take all information gathered from security devices like cameras and door readers and turn it into real-time actionable data. One problem this will address is the sheer volume of security notifications at a large enterprise. "Here at Microsoft, we have over a million alarm point events a week," says Anderson. At Microsoft all of the buildings have cafes, and often employees hold open doors for each other. It is not practical to respond to every door-held-open alarm simply due to the volume. However, Anderson says, if four door-held-open alarms happen in a row, at the front door, then at internal doors, then at an upper floor, that establishes a pattern where someone could be bringing something in. "So we want to teach the machine to look for those patterns that a human really couldn't see because of the volume," Anderson says. The system would then trigger an alert for the human operator to evaluate the situation.

Machine learning in video analytics could involve building heat maps showing normal activity in the space, and when the camera detects motion outside of that heat map an alarm is generated. In the future, facial recognition, which is already used in government settings, might be used in a corporation to detect persons of interest, such as terminated employees returning to campus, or perhaps an unauthorized person entering a secure area, even if they used a badge with the proper credentials to swipe in. "It's using all of these edge devices that every building has, cameras, card readers, and intrusion systems, and taking that data and putting it somewhere so that a machine can learn, and make it useful," says Anderson. "That's a huge step in the future."

Tech on deck

A security technology which is still finding its footing is using LED fixtures as a platform for surveillance and other security-related devices. The premise is that, as native digital appliances, LED fixtures can house any range of devices, such as cameras, occupancy sensors, wayfinding beacons, and on. "So now you can put a camera anywhere you want, any time," says Ray Bernard, president and principal consultant, Ray Bernard Consulting Services. A distributed security network enabled by the ceiling fixtures could be instrumental in locating individuals during an emergency, for example. The challenge is having the foresight to install the hardware at the beginning of a project so that the cost is minimized. A key question: "What is it that you want to put in when you're putting in intelligent LED lighting that maybe you won't be able to use right away, but will be there next year." This is particularly important given the longevity of LED fixtures.

One change Anderson expects to see sooner than later is the dissolution of the traditional security operations center in favor of a cloud-as-a-service model. A benefit will be that everyone on the security team responding to an event will be getting the same information in the same format. "Your edge security managers will be able to participate more in step," Anderson says.

Chapter Three

Future Possibilities

The core chapter throws lots of lights on innovations that are live as adopted in diverse sectors of human endeavour, there are no limits if only the visions can come alive, possibilities are enormous. Wastes converted to use , Applications spiced with Artificial intelligence help to solve complex processes.

There are possibilities of the future and they are as follows:

1) Large bank MPPT Charge controller

Large bank charge controller is not commercially available, my friend Engr Israel Omosagba, Meanwell Electric Limited have spent several months on this project,

382V charge controller, a single box. The practice is to deploy several MPPT controllers for >=100KVA Solution and connecting them in series. This happens to be the most critical component of a PV deployment and this attributes to failure of solar solutions. The prototype has been designed and implemented; it is now been tested at one of our sites in Abuja. This is a product for the future and proudly Nigerian.

2) Energy storing bricks

Scientists have found a way to store energy in the red bricks that are

used to build houses. Researchers led by Washington University in St Louis, in Missouri, US, have developed a method that can turn the cheap and widely available building material into "smart bricks" that can store energy like a battery.

Although the research is still in the proof-of-concept stage, the scientists claim that walls made of these bricks "could store a substantial amount of energy" and can "be recharged hundreds of thousands of times within an hour". The researchers developed a method to convert red bricks into a type

of energy storage device called a supercapacitor. This involved putting a conducting coating, known as Pedot, onto brick samples, which then seeped through the fired bricks' porous structure, converting them into "energy storing electrodes". Iron oxide, which is the red pigment in the bricks, helped with the process, the researchers said.

3) Robotic guide dogs

A student at Loughborough University has designed a "robotic guide dog" that will help support visually impaired people who are unable to house a real animal. The product, designed by Anthony Camu, replicates the functions of a guide dog as well as programming quick and safe routes to destinations using real-time data.

Theia, named after the titan goddess of sight, is a portable and concealable handheld device that guides users through outdoor environments and large indoor spaces with little input. Using a special control moment gyroscope (CMG), Theia moves users' hands and physically "leads" them – much like holding the brace of a guide dog.

The device is designed to process real-time online data, such as traffic density (pedestrians and cars) and weather, to guide users accurately and safely to their destinations. It will have a fail-safe procedure for high-risk scenarios, such as crossing busy roads – pushing the user back into a "manual mode", like using a cane.

4) Sweat powered smartwatches

   Engineers at the University of Glasgow have developed a new type of flexible supercapacitor, which stores energy, replacing the electrolytes found in conventional batteries with sweat.

   It can be fully charged with as little as 20 microlitres of fluid and is robust enough to survive 4,000 cycles of the types of flexes and bends it might encounter in use. The device works by coating polyester cellulose cloth in a thin layer of a polymer, which acts as the supercapacitor's electrode.

As the cloth absorbs its wearer's sweat, the positive and negative ions in the sweat interact with the polymer's surface, creating an electrochemical reaction which generates energy.

"Conventional batteries are cheaper and more plentiful than ever before but they are often built using unsustainable materials which are harmful to the environment," says Professor Ravinder Dahiya, head of the Bendable Electronics and Sensing Technologies (Best) group, based at the University of Glasgow's James Watt School of Engineering.

"That makes them challenging to dispose of safely and potentially harmful in wearable devices, where a broken battery could spill toxic fluids on to skin. "What we have been able to do for the first time is show that human sweat provides a real opportunity to do away with those toxic materials entirely, with excellent charging and discharging performance.

5) Self-healing 'living concrete'

Scientists have developed what they call living concrete by using sand, gel, and bacteria. Researchers said this building material has structural load-bearing function, is capable of self-healing and is more environmentally friendly than concrete – which is the second most-consumed material on Earth after water.

The team from the University of Colorado Boulder believe their work paves the way for future building structures that could "heal their own cracks, suck up dangerous toxins from the air or even glow on command".

6) Living robots

Tiny hybrid robots made using stem cells from frog embryos could one day be used to swim around human bodies to specific areas requiring medicine, or to gather microplastic in the oceans. "These are novel living machines," said Joshua Bongard, a computer scientist and robotics expert at the University of Vermont, who co-developed the millimetre-wide bots, known as xenobots.

"They're neither a traditional robot nor a known species of animal. It's a new class of artefact: a living, programmable organism.

7) Internet for everyone

We cannot seem to live without the internet (how else would you read sciencefocus.com?), but still only around half the world's population is connected. There are many reasons for this, including economic and social reasons, but for some the internet just is not accessible because they have no connection.

Google is slowly trying to solve the problem using helium balloons to beam the internet to inaccessible areas, while Facebook has abandoned plans to do the same using drones, which means companies like Hiber are stealing a march. They have taken a different approach by launching their own network of shoebox-sized microsatellites into low Earth orbit, which wake up a modem plugged into your computer or device when it flies over and delivers your data. Their satellites orbit the Earth 16 times a day and are already being used by organisations like The British Antarctic Survey to provide internet access to very extreme of our planet.

8) Heart monitoring T-shirt

Wearable sports bands that measure your heart rate are nothing new, but as numerous studies have shown, the accuracy can vary wildly (especially if you rely on them to count calories). In general, that is fine if you just want an idea of how hard you're working out, but for professionals, accuracy is everything.

Using a single lead ECG printed into the fabric, this new t-shirt from smart materials company KYMIRA will accurately measure heart beats and upload them to the cloud via Bluetooth. Once there, algorithms process the data to accurately detect irregular heartbeats such as arrhythmia heart beats, which could prove lifesaving.

And it is not just athletes who could benefit. "The possibilities this product offers both sportspeople and the general public is astonishing," says Tim Brownstone, CEO and founder of KYMIRA. "We envisage developing this product to be used for clinical applications to allow those who may already suffer with heart conditions enough warning of a heart attack."

9) Coffee power

London's coffee industry creates over 200,000 tonnes of waste every year, so what do we do with it? Entrepreneur Arthur Kay's big idea is to use his company, bio-bean, to turn 85 per cent of coffee waste into biofuels for heating buildings and powering transport.

"Anything's possible if you've got enough nerve." –J.K. Rowling

10)The AI scientist

Cut off a flatworm's head, and it'll grow a new one. Cut it in half, and you'll have two new worms. Fire some radiation at it, and it'll repair itself. Scientists have wanted to work out the mechanisms involved for some time, but the secret has eluded them. Enter an AI coded at Tufts University, Massachusetts. By analysing and simulating countless scenarios, the computer was able to solve the mystery of the flatworm's regeneration in just 42 hours. In the end it produced a comprehensive model of how the flatworm's genes allow it to regenerate.

Although humans still need to feed the AI with information, the machine in this experiment was able to create a new, abstract theory independently – a huge step towards the development of a conscious computer, and potentially a

"If you want something new, you have to stop doing something old." – Peter F. Drucker

landmark step in the way we carry out research.

11)Space balloon

If you want to take a trip into space, your quickest bet might be to take a balloon. The company World View Enterprises wants to send tourists into the stratosphere, 32km above Earth, on hot air balloons.

Technically 'space' is defined as 100km above sea level, but 32km is high enough to witness the curvature of the Earth, just as Felix Baumgartner did on his space jump. The balloon flew its first successful test flight in June, and the company will start selling tickets in 2016 – at the bargain price of just £75,000 per person!

12)Cancer-detecting 'smart needles'

A "smart needle" has been developed by scientists in the UK which could speed up cancer detection and diagnosis times.

Researchers believe the technology could be particularly helpful in diagnosing lymphoma, reducing patient anxiety as they await their results. At present, people with suspected lymphoma often have to provide a sample of cells, followed by a biopsy of the node to be carried out for a full diagnosis, a process which can be time consuming.

The new device uses a technique known as Raman spectroscopy to shine a low-power laser into the part of the body being inspected, with the potential to spot concerns within seconds, scientists from the University of Exeter say.

"The Raman smart needle can measure the molecular changes associated with disease in tissues and cells at the end of the needle," said professor Nick Stone, project lead, from the University of Exeter. "Provided we can reach a lump or bump of interest with the needle tip, we should be able to assess if it is healthy or not

13 **Car batteries that charge in 10 minutes**

Fast-charging of electric vehicles is seen as key to their take-up, so motorists can stop at a service station and fully charge their car in the time it takes to get a coffee and use the toilet – taking no longer than a conventional break. But rapid charging of lithium-ion batteries can degrade the batteries, researchers at Penn State University in the US say. This is because the flow of lithium particles known as ions from one electrode to another to charge the unit and hold the energy ready for use does not happen smoothly with rapid charging at lower temperatures.

However, they have now found that if the batteries could heat to 60°C for just 10 minutes and then rapidly cool again to ambient temperatures, lithium spikes would not form and heat damage would be avoided. The battery design they have come up with is self-heating, using a thin nickel foil which creates an electrical circuit that heats in less than 30 seconds to warm the inside of the battery. The rapid cooling that would be needed after the battery is charged would be done using the cooling system designed into the car. Their study, published in the journal Joule, showed they could fully charge an electrical vehicle in 10 minutes.

14 Self-driving trucks

We've almost got used to the idea of driverless cars before we've even seen one on the roads. The truth is, you might well see a lot more driverless trucks – after all, logistics make the world go round. They'll be cheaper to run than regular rigs, driving more smoothly and so using less fuel. Computers never get tired or need comfort breaks, so they'll run longer routes. And they could drive in convoys, nose-to-tail, to minimise wind resistance.

Companies like Mercedes and Peloton are already exploring these possibilities, and if the promised gains materialise, freight companies could upgrade entire fleets overnight. On the downside, it could put drivers instantly out of work, and even staff at the truck stops set up to service them, but many companies have said the trucks will still need a human passenger to ensure their cargo is safe.

15 **Floating farms**

The UN predicts there will be two billion more people in the world by 2050, creating a demand for 70 per cent more food. By that time, 80 per cent of us will be living in cities, and most food we eat in urban areas is brought in. So farms moored on the sea or inland lakes close to cities would certainly reduce food miles. But how would they work? A new design by architect Javier Ponce of Forward-Thinking Architecture shows a 24m-tall, three-tiered structure with solar panels on top to provide energy. The middle tier grows a variety of veg over an area of 51,000m2, using not soil but nutrients in liquid. These nutrients and plant matter would drop into the bottom layer to feed fish, which are farmed in an enclosed space.

A single Smart Floating Farm measuring 350 x 200m would produce an estimated 8.1 tonnes of vegetables and 1.7 tonnes of fish a year. The units are designed to bolt together, which is handy since we'll need a lot of them: Dubai, for instance, imports 11,000 tonnes of fruit and veg every day.

There are no magic formulas for innovation. However, with the huge number of changes that have taken place in the modern-day world, you must never think that innovation will be following the same rules that have been in place for decades. The context we live and work in affects us. It affects the way we work (otherwise, something is not going right), the way we relate with others, the way we dress, and, in general, the way we live. Of course, this context must necessarily affect the way we approach innovation. And if you think our environment has changed a lot lately, just wait and see what we are headed for. Start getting ready to see a great deal of change, much more radical, and much faster. If you or your company are not yet aware of this, you had better start working toward that goal.

www.ingramcontent.com/pod-product-compliance
Lightning Source LLC
Chambersburg PA
CBHW070549220526
45467CB00003B/1137

* 9 7 9 8 6 9 7 9 4 6 8 3 1 *